Fuse foundation

A Comprehensive Guide to Fusion Anatomy

Dr Joe smith

Contents

Chapter1 3
introduction to anatamony of fuse 3
Chapter2 13
Construction and working of a fuse 13
chapter3 26
how to select a proper rating size of a fuse 26
The end 35

Chapter 1

introduction to anatamony of fuse

Anatomy of fuse is an essential topic in the field of electrical engineering, which deals with the study of electrical circuits and the components that make them up. A fuse is one of the crucial components in an electric circuit, serving as a protective device that prevents the circuit from getting damaged due to overloading or short circuits. In simple terms, a fuse is a small, easily replaceable device that is connected in series with a circuit, designed to break the connection when there is an excessive current flow. The basic anatomy of fuse comprises of two essential parts, a fuse element or wire and a fuse holder or casing. The fuse element is usually a thin metal strip that is selected based on the expected current flow in the circuit. The fuse holder, on the other hand, is a non-conducting material that encloses the fuse element and provides support and insulation to it. To understand the anatomy of fuse better, it is essential to know how it works. When current passes through an electrical circuit, it flows from a source, such as a battery or power supply, through various components and back to the source. This current creates heat due to the resistance of the circuit's components. If the current exceeds the rated capacity of the fuse element, the heat generated will cause the fuse element to melt or break, thus opening the circuit and interrupting the current flow. The fuse holder plays a critical role in this process by providing a safe and secure housing for the fuse element. It is usually made up of a material such as porcelain, ceramic, or plastic, which is known for its high insulation properties. This helps prevent accidental electric shocks and provides protection against fire hazards. Apart from this, the fuse holder also acts as a support for the fuseAnatomy of fuse refers to the physical structure and composition of a device that is used for protecting electrical circuits from damage due to overcurrent. It is a crucial component of any electrical system and plays a vital role in ensuring the safe and efficient functioning of various electrical devices.

A fuse is a safety device that is designed to melt or break a circuit when the current flowing through it exceeds a certain limit. This current limit, also known as the "rated current," is determined by the fuse's design and size. Once the fuse is "blown," it prevents the flow of excessive current, thus protecting the circuit and the device it is connected to from damage. The history of fuses can be traced back to the 1800s when they were first used in the early electric telegraph systems. However, with the advancement of technology and the widespread use of electricity, fuses have evolved significantly over the years, both in terms of their design and composition. One of the most important aspects of fuse anatomy is its construction. A typical fuse consists of a metal strip or wire, known as the "element," which is surrounded by a non-conductive material such as porcelain, glass, or ceramic, known as the "body." The body of the fuse is often cylindrical in shape and can be of different sizes, depending on the application. The element of a fuse is usually made of materials such as copper, silver, or aluminum, which are good conductors of electricity. The choice of material for the element is crucial as it needs to have a low melting point and be able to carry the rated current without overheating. Copper is the most commonly used material for elements in fuses as it has a relatively low melting point and high thermal conductivity. Apart from the element and the body, a fuse also consists of two terminals, one at each end, through which it is connected to the circuit. These terminals are made of materials that can withstand high temperatures and have good conductivity, such as copper or brass. One of the key components of fuse anatomy is the "melting point." This refers to the temperature at which the element inside the fuse will melt or break, thus interrupting the flow of current. The melting point of a fuse is determined by its material and size, and it is crucial to choose the right type of fuse with the appropriate melting point for a specific circuit. Another critical feature of fuse anatomy is the "breaking capacity." This refers to the maximum amount of current that a fuse can safely interrupt without causing any damage. The breaking capacity of a fuse is determined by the material and size of the element, as well as the design of the fuse body. Fuses also have a "time-current characteristic" that defines the relationship between the time it takes for the fuse to blow and the amount of current flowing through it. This characteristic is crucial in determining the suitability of a fuse for a particular application. For instance, in highly sensitive electrical circuits, a fast-acting fuse with a lower time-current characteristic is preferred, while in less critical circuits, a slow-blow fuse with a higher time-current characteristic may be suitable. Apart from these design features, fuses also have different types of designs, such as cartridge fuses, plug fuses, and blade fuses. These designs vary in terms of their shape, size, and mounting, and are chosen based on the specific application and voltage level. Cartridge fuses, also known as cylindrical fuses, are the most common type of fuse used in industrial and commercial applications. They are made of a thin-walled glass or ceramic tube and have metal end caps through which the terminals are connected. Cartridge fuses are available in a variety of sizes and ratings and are used in a wide range of electrical equipment. Plug fuses, on the other hand, are designed for residential and light commercial applications and are typically used to protect household appliances and lower voltage circuits. These fuses have a cylindrical base with a screw-type adapter and a round, glass or ceramic body. Plug fuses are available in various sizes and ratings and are commonly used in household electrical outlets. Blade fuses, also known as spade fuses, are relatively new designs and have become popular in modern vehicles and equipment. These fuses have a plastic body with two terminals at each end, and the element is housed within the body. Blade fuses are smaller in size compared to cartridge fuses and are available in different colors to identify their ratings easily. In addition to their design and construction, the materials used in fuses have also evolved over the years. While the most commonly used material for elements is copper, newer technologies have led to the development of fuses with elements made of silver, which has a lower melting point and can handle higher currents. Furthermore, the use of "semiconductor" materials, such as zinc oxide, has also become popular in modern fuses. These materials have a much higher resistance compared to traditional materials, allowing them to act as "overload indicators" by creating a voltage drop when current exceeds the rated limit. This feature is particularly useful in monitoring and troubleshooting electrical systems.

Chapter 2

Construction and working of a fuse

The construction and working of a fuse are vital to understanding its importance and how it provides protection to electrical circuits. Construction: Fuses come in various shapes and sizes depending on their application, but they all have a similar construction. The most common type of fuse is a cartridge fuse, which consists of a cylindrical body made of ceramic, glass, or plastic material. Inside the body, there is a thin wire called the fuse element, which is usually made of a low-melting-point alloy or copper. The fuse element is connected to two metallic end caps at the bottom of the body, which are used to connect the fuse to the circuit. Working: Fuses work based on the principle of electrical resistance and the relationship between current and temperature. When current is passed through a fuse, the fuse element heats up due to its resistance, and the temperature increases proportionally to the amount of current flowing through it. As the temperature rises, the fuse element reaches its melting point, and the wire melts, breaking the circuit and cutting off the flow of current. The rating of a fuse is determined by the current-carrying capacity of the fuse element, and it is usually marked on the body of the fuse. When the current flowing through the circuit exceeds the rated value of the fuse, the fuse element will melt, and the fuse will blow, indicating that it needs to be replaced. Types of fuses: 1. Fast-acting fuse: This type of fuse is designed to blow quickly in case of a sudden surge of current. It is commonly used in electronic circuits to protect sensitive components from damage. 2. Slow-blow fuse: Slow-blow fuses are used in circuits that require a higher start-up current or a higher level of current for a short time. They are designed to withstand this initial higher current before blowing. 3. Thermal fuse: A thermal fuse is a type of fuse that is activated when the temperature exceeds a certain threshold. It is commonly used in household appliances such as hair dryers and toasters. 4. Resettable fuse: As the name suggests, a resettable fuse can be reset after it blows. It uses a polymeric positive temperature coefficient (PPTC) material that has a high resistance when cool but becomes conductive when heated. When the fuse blows, the PPTC material heats up and increases its resistance, cutting off the flow of current. Once the current is removed, the material cools down and returns to its initial high resistance state. 5. Automotive fuse: These types of fuses are used in vehicles and are designed to tolerate high levels of vibration and shock. They are available in different ratings and can be easily replaced without the need for special tools. Importance of fuses: Fuses play a crucial role in protecting electrical circuits from overcurrent, which can cause damage to the circuit and its components. They also act as a safety measure to prevent fires and electrocution. In addition, fuses are relatively inexpensive and easy to replace, making them an ideal choice for circuit protection. Fuses also serve as an early warning system for potential electrical faults. When a fuse blows, it is an indication that there is an issue with the circuit, and it needs to be investigated. This helps in preventing further damage or potential hazards. Limitations of fuses: One of the main limitations of fuses is that they are not reusable and need to be replaced after blowing. This can be costly and time-consuming, especially in large electrical systems. In addition, fuses can only provide overcurrent protection, and they cannot protect against other types of faults, such as short circuits. Another limitation is the time delay in the fuse blowing. In certain situations, the fuse may not blow fast enough, and this can result in damage to the circuit or its components. This is why it is important to select the right type and rating of fuse for a specific application.

characteristics of a fuse

Composition and Construction: Fuses are typically made of a conductive material, such as copper, silver or aluminum, which is housed in a non-conductive casing, such as glass, ceramic, or plastic. The fuse's conductive element is designed to melt and break the circuit if an excessive amount of current flows through it. The casing provides physical protection and insulation for the fuse and helps in holding the conductive element in place. 2. Current Rating: One of the primary characteristics of a fuse is its current rating, which determines how much current the fuse can handle without melting. This rating is usually given in amperes (A) and is labeled on the fuse or its casing. Fuses are available in a wide range of current ratings to suit different applications, and it is essential to select the right fuse with an appropriate current rating for the specific circuit. 3. Time-Current Characteristics: The time-current characteristic of a fuse determines the amount of time it takes for the fuse to blow at different levels of current. A slow-blow fuse, also known as time-delay fuse, has a gradual time-current curve and can withstand a brief overload without blowing. These types of fuses are commonly used in applications where the starting current of the equipment or motor is significantly higher than its normal operating current. On the other hand, a fast-blow fuse has a steep time-current curve and blows almost instantly when the current reaches its rated value. These fuses are ideal for protecting delicate electronic components that are susceptible to damage from excessive current. 4. Voltage Rating: The voltage rating of a fuse is the maximum voltage that the fuse can handle without breaking down or arcing. It is an essential characteristic because if the voltage exceeds the fuse's rating, it may create a high-energy arc that can damage the fuse or cause a fire hazard. Therefore, it is crucial to select a fuse with a voltage rating that is equal to or higher than the circuit's operating voltage. 5. Interrupting Rating: The interrupting rating, also known as short-circuit rating, is the maximum amount of current that a fuse can safely interrupt without causing harm. When a circuit experiences a short circuit or excessive overload, the fuse must be able to break the circuit and stop the flow of current. If a fuse with a low interrupting rating is used in a circuit with a high fault current, it may result in the fuse exploding or causing a fire. Therefore, it is essential to select a fuse with an appropriate interrupting rating for the specific application. 6. Operating Temperature: Fuses are designed to work within a specified temperature range, and their operating characteristics can be affected if the temperature falls out of this range. For example, if a fuse is exposed to a higher temperature, it may melt and break the circuit at a lower current than its rated value. Similarly, at a lower temperature, the fuse may take longer to blow, which can result in damage to the equipment or the fuse itself. It is essential to choose a fuse with an appropriate temperature rating that can withstand the conditions of its intended application. 7. Sensitivity: Fuses have a specific level of sensitivity to current flow, which is determined by the type of conductive material used in them. Fuses made of copper have a lower resistance to current flow, making them more sensitive than those made of aluminum. The sensitivity of a fuse is also affected by its size, with smaller fuses being more sensitive than larger ones. This characteristic is crucial in applications where even a small amount of overcurrent can cause damage, such as in electronic circuits. 8. Repeatability: Certain types of fuses, such as thermal or PTC (Positive Temperature Coefficient) fuses, have the ability to reset and continue functioning after they have blown. This characteristic makes them suitable for applications where overcurrent may occur momentarily, such as in a motor starting up. Once the abnormal current is cleared, these fuses cool down and become functional again, thus eliminating the need for replacement. 9. Degradation: Fuses are subject to degradation due to factors such as high temperature, excessive current, and aging. Repeated exposure to these conditions can impact the fuse's ability to operate at its specified level and may result in a false triggering or failure to blow when needed. It is essential to regularly inspect and replace fuses if any signs of degradation are noticed to ensure their proper functioning. 10. Single-use: The most common type of fuse is a single-use fuse, meaning that it needs to be replaced once it has blown. This characteristic makes it essential to keep spare fuses on hand in case of emergencies. However, there are also fuses available that can be reset and used multiple times, such as those used in vehicles.

chapter3

how to select a proper rating size of a fuse

There are various factors to consider when selecting a proper fuse size, such as the type of circuit, the expected load, and the ambient temperature. In this 1000-word essay, we will discuss how to select the right rating size of a fuse for your specific electrical system. Understand the Basics of Fuses Before diving into the process of selecting a proper fuse size, it is essential to understand the basics of fuses. A fuse is a small, thin, and glass or ceramic tube with a thin strip of metal inside. When excessive current flows through the circuit, the metal strip melts, breaking the circuit and stopping the flow of electricity. This process protects the circuit and its components from damage. Fuses are designed with a predetermined rating known as "ampere" or "amp" rating. It indicates the maximum amount of current the fuse can safely handle before blowing out. Amp ratings of fuses can range from fractions of an amp to hundreds of amps, depending on the intended purpose and application. Consider the Type of Circuit The first factor to consider when selecting a proper fuse size is the type of circuit it will be used in. The two main types of circuits are alternating current (AC) and direct current (DC). AC circuits are commonly used in homes and businesses, while DC circuits are typically used in electronic devices and vehicles. The type of circuit will dictate the type of fuse to be used. For AC circuits, standard type fuses such as cartridge fuses or plug fuses are commonly used. These fuses have a higher voltage rating and are designed to break the circuit when a higher current flows through them. For DC circuits, special type fuses such as blade or semiconductor fuses are more suitable. These fuses have a lower voltage rating and are designed to break the circuit when a lower current flows through them. Determine the Expected Load The expected load refers to the amount of current that the circuit is designed to carry. It is important to determine the maximum expected load before selecting a fuse size, as using a fuse with a higher rating can result in inadequate circuit protection, while using a fuse with a lower rating can cause frequent and unnecessary tripping. To determine the expected load, you will need to calculate the maximum current that the circuit will draw. This can be done by adding the current ratings of all the devices connected in the circuit. It is always recommended to leave some margin, typically 25%, to account for any unexpected current spikes. Consider the Ambient Temperature The temperature of the environment in which the fuse will be used is another crucial factor to consider when selecting a fuse size. Fuses are designed to operate within a specific temperature range, known as the "operating temperature." If the temperature of the environment exceeds this range, the fuse can blow unnecessarily, causing inconvenience and extra expenses. On the other hand, if the temperature is too low, the fuse may not blow when necessary, resulting in damage to the circuit. To determine the correct fuse size, it is essential to calculate the derating factor, which is the reduction in the fuse's amp rating based on the ambient temperature. This information can typically be found in the manufacturer's datasheet. Consider the Voltage Rating The voltage rating of a fuse is also a crucial factor to consider when selecting a proper fuse size. It represents the maximum voltage that the fuse can safely handle before breaking the circuit. Using a fuse with a lower voltage rating can result in arcing and cause fire hazards, while using a fuse with a higher voltage rating may not provide proper circuit protection. The voltage rating of a fuse should always be equal to or higher than the maximum voltage expected in the circuit. It is best to consult a professional or adhere to the manufacturer's recommendations for selecting the correct voltage rating for your specific application. Check the Time-Current Characteristics Another important consideration when selecting a fuse size is the time-current characteristics, also known as the time-delay or time lag characteristics. These characteristics determine the amount of time it takes for the fuse to blow when exposed to a specific amount of current.

Some circuits may require a fast-acting fuse that can blow quickly when overloaded, while others may require a time-delay fuse to accommodate temporary current spikes. It is crucial to check the time-current characteristics of a fuse and make sure it is suitable for the specific circuit it will be used in. Consult a Professional or Use an Online Calculator Selecting the proper fuse size can be a complex process, and it is always best to consult a professional, such as an electrician or an engineer, to determine the correct fuse for your specific application. They can take into account all the aforementioned factors and provide a reliable and safe solution. If you prefer to do it yourself, there are various online calculators and tools available that can assist you in selecting the proper fuse size. These calculators consider all the necessary factors and provide you with a recommended fuse size based on your inputs. It is important to note that these tools are meant to be used as a guide and should not replace professional advice. Consider Different Brands and Types of Fuses There are countless brands and types of fuses available on the market, each with its own unique characteristics and ratings. It is essential to make an informed decision by comparing different brands and types of fuses and selecting the one that best suits your specific needs. Some factors to consider when comparing different fuses include their voltage rating, current rating, time-current characteristics, and price. It is also recommended to purchase fuses from reputable brands and authorized dealers to ensure their quality and reliability. Regular Inspection and Maintenance Selecting the proper fuse size is only the first step in ensuring proper circuit protection. It is essential to regularly inspect and maintain your fuses to ensure their reliability and effectiveness. Factors such as aging, dust, and corrosion can affect the performance of fuses, and it is important to replace them when necessary to prevent electrical hazards.

The end

www.ingramcontent.com/pod-product-compliance
Lightning Source LLC
Chambersburg PA
CBHW071202240526
45470CB00017B/1240